THE MECHANICAL ACTION OF LIGHT

By
WILLIAM CROOKES, F. R. S.

ISBN-13:
978-1727163094

ISBN-10:
1727163095

SOME experiments illustrating the mechanical action of light, which I have recently exhibited before the Fellows of the Royal Society, having attracted considerable attention, I propose to give here a description of some of the instruments which my researches have enabled me to construct. But, to render the subject more intelligible, it will be necessary to give a brief outline of the researches which I have been carrying on for the last three or four years, so that the reader may see the gradual steps which have led up to the full proof that radiation is a motive power.

The experiments were first suggested by some observations made when weighing heavy pieces of glass apparatus in a chemical balance, inclosed in an iron case from which the air could be exhausted. When the substance weighed was of a temperature higher than that of the surrounding air and the weights, there appeared to be a variation of the force of gravitation. Experiments were thereupon instituted to render the action more sensible and to eliminate sources of error.[1]

My first experiments were performed with apparatus made on the principle of the balance. An exceedingly fine and light arm was delicately suspended in a glass tube by a double-pointed needle; and at the ends were affixed balls of various materials. Among the substances thus experimented on I may mention pith, glass, charcoal, wood, ivory, cork, selenium, platinum, silver, aluminium, magnesium, and various other metals.

The most delicate apparatus for general experiment was made with a straw beam having pith masses at the end. The general appearance of the apparatus is shown in Fig. 1.

A is the tube belonging to the Sprengel pump.[2] *B* is the desiccator, full of glass beads moistened with sulphuric acid. *C* is the tube containing the straw balance with pith ends: it is drawn out to a contracted neck at the end connected with the pump, so as to readily admit of being sealed off at any stage of the exhaustion. *D* is the pump-gauge, and *E* is the barometer.

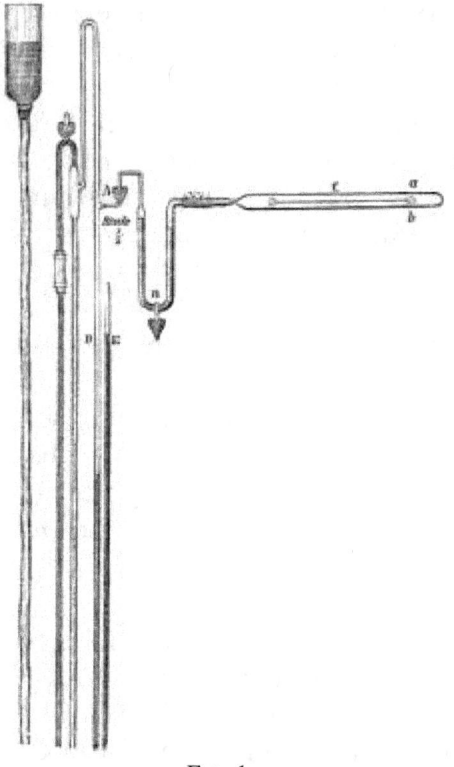

Fig. 1.

The whole being fitted up as here shown, and the apparatus being full of air to begin with, I passed a spirit-flame across the lower part of the tube at *b,* observing the movement by a low-power micrometer: the pith-ball (*a h*) descended slightly, and then immediately rose to considerably above its original position. It seemed as if the true action of the heat was one of attraction, instantly overcome by ascending currents of air. A hot metal or glass rod and a tube of hot water applied beneath the pith-ball at *b* produced the same effect as the flame; when applied above at a they produced a slight rising of the ball. The same effects take place when the hot body is a])plied to the other end of the balanced beam. In these cases air-currents are sufficient to explain the rising of the ball under the influence of heat.

In order to apply the heat in a more regular manner, a thermometer was inserted in a glass tube, having at its extremity a glass bulb about one and a half inch diameter; it was filled with water and then sealed up (*see* Fig. 2). This was arranged on a revolving stand, so that by means of a cord I could bring it to the desired position without moving the eye from the micrometer. The water was kept heated to 70° C, the temperature of the laboratory being about 15° C.

FIG. 2.

The barometer being at 767 millimetres and the gauge at zero, the hot bulb was placed beneath the pith-ball at *b*. The ball rose rapidly. The source of heat was then removed, and as soon as equilibrium was restored I placed the hot-water bulb above the pith-ball at *a,* when it rose again—more slowly, however, than when the heat was applied beneath it.

The pump was then set to work; and when the gauge was 147 millimetres below the barometer, the experiment was tried again: a similar result, only more feeble, was obtained. The exhaustion was continued, stopping the pump from time to time to observe the effect of heat, when it was seen that the effect of the hot body regularly diminished as the rarefaction increased, until, when the gauge was about twelve millimetres below the barometer, the action of the hot body was scarcely noticeable. At ten millimetres below it was still less; while when there was only a difference of seven millimetres between the barometer and the gauge, neither the hot-water bulb, the hot rod, nor the spirit-flame, caused the ball to move in an appreciable degree.

The inference was almost irresistible that the rising of the pith was only due to currents of air, and that at this near approach to a vacuum the residual air was too highly rarefied to have power in its rising to overcome the inertia of the straw beam and the pith-balls. A more delicate instrument would doubtless show traces of movement at a still nearer approach to a vacuum; but it seemed evident that when the last trace of air had been removed from the tube surrounding the balance—when the balance was suspended in empty space only—the pith-ball would remain motionless, wherever the hot body was applied to it.

I continued exhausting. On next applying heat underneath, the result showed that I was far from having discovered the law governing these phenomena; the pith-ball rose steadily, and without that hesitation which had been observed at lower rarefactions. With the gauge three millimetres below the barometer, the ascension of the pith when a hot body was placed beneath it was equal to what it had been in air of ordinary density; while with the gauge and barometer level its upward movements were not only sharper than they had been in air, but they took place under the influence of far less heat—the finger, for example, instantly repelling the ball to its fullest extent.

To verify these unexpected results, air was gradually let into the apparatus, and observations were taken as the gauge sank. The same effects were produced in inverse order, the point of neutrality being when the gauge was about seven millimetres below a vacuum.

A piece of ice produced exactly the opposite effect to a hot body.

The presence of air having so marked an influence on the action of heat, an apparatus was fitted up in which the source of heat (a platinum spiral rendered incandescent by electricity) was inside the vacuum-tube instead of outside it as before; and the pith-balls of the former apparatus were replaced by brass balls. By careful manipulation and turning the tube round, I could place the equipoised brass ball either over, under, or at the side of the source of heat. With this apparatus I tried many experiments, to ascertain more about the behavior of the balance during the progress of the exhaustion, both below and above the point of no action, and also to ascertain the pressure corresponding with this critical point.

In one experiment, which is described in detail in my paper on this subject before the Royal Society,[3] the pump was worked until the gauge had risen to within five millimetres of the barometric height. On arranging the ball above the spiral, and making contact with the battery, the attraction was still strong, drawing the ball downward a distance of two millimetres. The pump continuing to work, the gauge rose until it was within one millimetre of the barometer. The attraction of the hot spiral for the ball was still evident, drawing it down when placed below it, and up when placed above it. The movement, however, was much less decided than before; and, in spite of previous experience, the inference was very strong that the attraction would gradually diminish until the vacuum was absolute, and that then, and not till then, the neutral point would be reached. Within one millimetre of a vacuum there appeared to be no room for a change of sign.

The gauge rose until there was only half a millimetre between it and the barometer. The metallic hammering heard when the rarefaction is close upon a vacuum commenced, and the falling mercury only occasionally took down a bubble of air. On turning on the battery-current, there was the faintest possible movement of the brass ball (toward the spiral) in the direction of attraction.

The working of the pump was continued. On next making contact with the battery, no movement could be detected. The red-hot spiral neither attracted nor repelled. I had arrived at the critical point. On looking at the gauge I saw it was level with the barometer.

The pump was now kept at full work for an hour. The gauge did not rise perceptibly, but the metallic hammering sound increased in sharpness, and I could see that a bubble or two of air had been carried down. On igniting the spiral, I saw that the neutral point had been passed. The sign had changed, and the action was one of faint but unmistakable *repulsion.* The pump was still kept going, and an observation was taken, from time to time, during several hours. The repulsion continued to increase. The tubes of the pump were now washed out with oil of vitriol,[4] and the working was continued for an hour.

The action of the incandescent spiral was now found to be energetically *repellent,* whether it was placed above or below the brass ball. The fingers-exerted a repellent action, as did also a warm glass rod, a spirit-flame, and a piece of hot copper.

In order to decide once for all whether these actions really were due to air-currents, a form of apparatus was fitted up which—while it would settle the question indisputably—would at the same time be likely to afford information of much interest.

By chemical means I obtained in an apparatus a vacuum so nearly perfect that it would not carry a current from a Ruhmkorff's coil when connected with platinum wires sealed into the tube. In such a vacuum the repulsion by heat was still found to be decided and energetic.

I next tried experiments in which the rays of the sun, and then the different portions of the solar spectrum, were projected on to the delicately-suspended pith-ball balance. *In vacuo* the repulsion by a beam of sunlight is so strong as to cause danger to the apparatus, and resembles that which would be produced by the physical impact of a material body.

A simpler form of the apparatus for exhibiting the phenomena of attraction in air and repulsion in a vacuum consists of a long glass tube, *a b* (Fig. 3), with a globe, *c,* at one end. A light index of pith, *d e,* is suspended in this globe by means of a cocoon fibre.

When the apparatus is full of air at ordinary pressure, a ray of heat or light falling on one of the extremities of the bar of pith gives a movement indicating attraction. When the apparatus is exhausted until the barometric gauge shows a depression of twelve millimetres below the barometer, neither attraction nor repulsion results when radiant light or heat falls on the pith, but, when the vacuum is as good

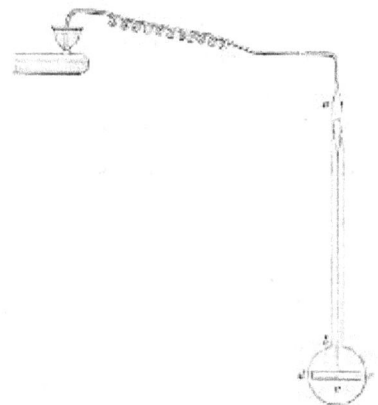

Fig. 3.

as the pump will produce, strong repulsion is shown when radiation is allowed to fall on one end of the index. An apparatus of this kind, constructed with the proper precautions, and sealed off when the vacuum is perfect, is so sensitive to heat that a touch with the finger on a part of the globe near one extremity of the pith will drive the index round over 90, while it follows a piece of ice as a needle follows a magnet. With a large bulb, very well exhausted, and containing a suspended bar of pith, a somewhat striking effect is produced when a lighted candle is placed about two inches from the globe. The pith-bar commences to oscillate to and fro, the swing gradually increasing in amplitude until the dead-centre is passed over, when several complete revolutions are made. The torsion of the suspending fibre now otters resistance to the revolutions, and the bar commences to turn in the opposite direction. This movement is kept up with great energy and regularity as long as the candle burns.

For more accurate experiments I prefer making the apparatus differently. Fig. 4 represents the best form: $a\ b$ is a glass tube, to which is fused at right angles another narrower tube, $c\ d;$ the vertical tube is slightly contracted at $e,$ so as to prevent the solid stopper *d*—which just fits the bore of the tube—from falling down. The lower end of the stopper, *d e,* is drawn out to a point; and to this is cemented

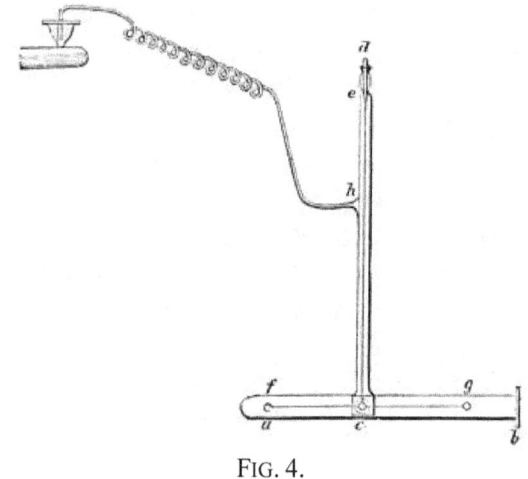

FIG. 4.

a fine glass thread, about 0.001 inch diameter, or less, according to the torsion required.[5]

At the lower end of the glass thread an aluminium stirrup and a concave glass mirror are cemented, the stirrup being so arranged that it will hold a beam, *f g,* having masses of any desired material at the extremities. At *c* in the horizontal tube is a plate-glass window cemented on to the tube.

At *b* is also a piece of plate-glass cemented on. Exhaustion is effected through a branch-tube, *h,* projecting from the side of the upright tube. This is sealed by fusion to the spiral tube of the pump. The stopper *d e* and the glass plates *c* and *b* are well fastened with a cement of resin and bees'-wax.

The advantage of a glass-thread suspension is that the beam always comes back to its original position.

An instrument of this sort, perfectly exhausted and then sealed off, is shown at work in Fig. 5. It has pith-plates at the extremities of the torsion-beam. A ray of light from the lamp is thrown on to the central mirror, and thence reflected on to the graduated scale. The approach of a finger to either extremity of the beam causes the luminous index to travel several inches, slowing repulsion. A piece of ice brought near causes the spot of light to travel as much in the opposite direction. In order to insure the luminous index coming accurately back to zero, extreme precautions must be taken to keep all extraneous radiation from acting on the torsion-balance. The whole apparatus is closely packed round with a layer of cotton-wool about six inches thick, and outside this is arranged a double row of

Winchester quart-bottles, filled with water, spaces only being left for the radiation to fall on the balance and for the index ray of light to get to and from the mirror.

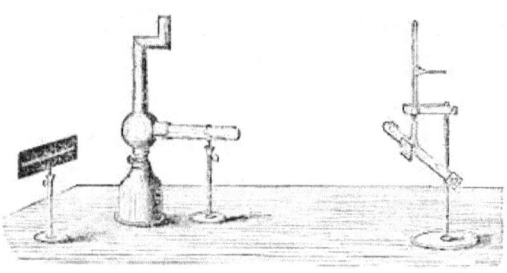

Fig. 5.

However much the results may vary when the vacuum is imperfect, with an apparatus of this kind they always agree among themselves when the residual gas is reduced to the minimum possible; and it is of no consequence what this residual gas is. Thus, starting with the apparatus full of various vapors and gases, such as air, carbonic acid, water, iodine, hydrogen, ammonia, etc., there is not found, at the highest rarefaction, any difference in the results which can be traced to the residual gas. A hydrogen vacuum appears the same as a water or an iodine vacuum.

The neutral point for a thin surface of pith being low, and that for a moderately thick piece of platinum being high, it follows that at a rarefaction intermediate between these two points pith will be repelled, and that platinum will be attracted by the same beam of radiation. This has been proved experimentally. An apparatus showing simultaneous attraction and repulsion by the same ray of light is illustrated in Fig. 6.

The pieces fg on the end of one beam consist of platinum-foil exposing a square centimetre of surface, while the extremities $f'g'$ on the other beam consist of pith-plates of the same size. A wide beam of radiation thrown in the centre of the tube on to the plates gf' causes g to be attracted and f' to be repelled, as shown by the light reflected from the mirrors, cc'. The atmospheric pressure in the apparatus is equal to about forty millimetres of mercury.

In a torsion-apparatus similar to the one shown in Figs. 4 and 5 I have submitted variously-colored disks to the action of the different rays of the spectrum. The . most striking results, as yet, have been obtained when the different rays of the spectrum were thrown on white and on black surfaces. The result was to show a decided difference between the actions of light and of radiant heat. At the highest exhaustions dark heat from boiling water acts almost equally on white pith and on pith coated with lamp-black, repelling either with about the same force. The action of the luminous rays, however, is different. These repel the black surface more energetically than they do the white surface, and, consequently, if in such an apparatus as is

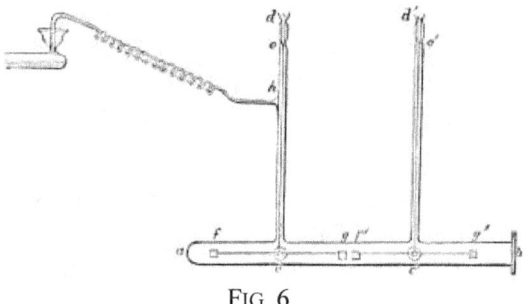

FIG. 6.

shown at Fig. 4, one disk of pith is white and the other is black, an exposure of both of them to light of the same intensity will cause the torsion-thread to twist round, owing to the difference of repulsion exerted on the black and the white surface. If, in the bulb-apparatus shown in Fig. 3, the halves of the pith-bar are alternately white and lamp-blacked, this differential action will produce rapid rotation in one direction, which keeps up until stopped by the torsion of the suspending fibre.

Taking advantage of this fact I have constructed an instrument which I have called the Radiometer, shown in section and plan at Figs. 7 and 8. It consists of four arms, of some light material, suspended on a hard steel point resting in a jewel-cup, so that the arms are able to revolve horizontally upon the centre pivot, in the same manner as the arms of Dr. Robinson's anemometer revolve. To the extremity of each arm is fastened a thin disk of pith, white on one side and lamp-blacked on the other, the black surfaces of all the disks facing the same way. The whole is inclosed in a thin glass globe, which is then exhausted to the highest attainable point and hermetically sealed.

☐The arms of this instrument rotate with more or less velocity under the action of radiation, the rapidity of revolution being directly proportional to the intensity of the incident rays. Placed in the sun,

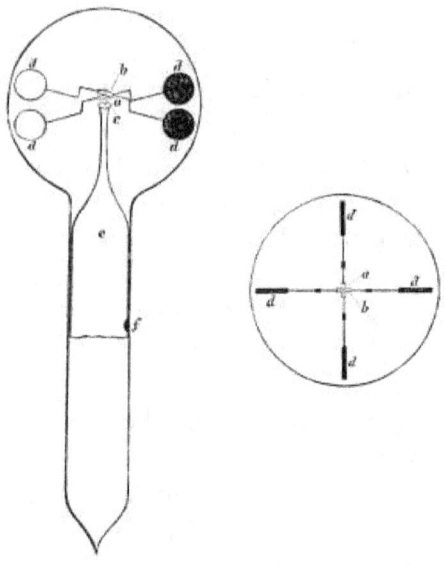

FIGS. 7 and 8.—*a*, a very fine needle-point; *b*, two pieces of straw; *c*, jewel-cup; *d, d, d, d*, four pith-disks, blackened on one side. The arms between the straw in the centre and the disks are bent glass-fibres; *e*, glass support holding cup; *f*, cement to keep the support *c* in its place.

or exposed to the light of burning magnesium, the rapidity is so great that the separate disks are lost in a circle of light. Exposed to a candle twenty inches off another instrument gave one revolution in 182 seconds; with the same candle placed at a distance of ten inches off the result is one revolution in 45 seconds; and at five inches off one revolution was given in 11 seconds. Thus it is seen that the mechanical action of radiation is inversely proportional to the square of the distance. At the same distance two candles give exactly double, and three candles give three times the velocity given by one candle, and so on up to twenty-four candles. A small radiometer was found to revolve at the velocities shown in the following table, when exposed to the radiation of a standard candle five inches off:

Time required for One Revolution.

Source of Radiation.					Time in Seconds.
1 candle, 5 inches off, behind			green	glass	40
"	5	"	"	blue	38
"	5	"	"	purple	28
"	5	"	"	orange	26
"	5	"	"	yellow	21

	" 5 "		" light-red □.□.□.	□"□.□. 20 □.□.

In diffused daylight the velocity was one revolution in from 1.7 seconds to 2.3 seconds, according to the intensity of the incident rays. In full sunshine, at 10 A. M., it revolved once in 0.3 second, and at 2 P. M. once in 0.25 second.

When heat is cut off by allowing the radiation to pass through a thick plate of alum, the velocity of rotation is somewhat slower, and when only dark heat is allowed to fall on the arms (as from a vessel of boiling water) no rotation whatever is produced.

In all respects, therefore, it is seen that the radiometer gives indications in strict accordance with theory.

Several radiometers, of various constructions as regards details, but all depending on the above-named discovery, have been exhibited at the Royal Society, where their novelty and unexpected indications excited a considerable amount of interest.

This form of instrument is of too recent a construction for me to be able to do more than draw brief attention to a few of the many uses for which it is applicable.

By timing the revolutions of the instrument when exposed direct to a source of light—a candle, for instance—the total radiation is measured. If a screen of alum is now interposed, the influence of heat is almost entirely cut off, the velocity becomes proportionately less, and the instrument becomes a photometer. By its means photometry becomes much simplified; flames the most diverse may readily be compared between themselves or with other sources of light; a "standard candle" can now be defined as one which at x inches off causes the radiometer to perform y revolutions per minute, the values of x and y having previously been determined by comparison with some ascertained standard; and the statement that a gas-light is equal to so many candles may, with more accuracy, be replaced by saying that it produces so many revolutions.

To photographers the radiometer will be invaluable. As it will revolve behind the orange-colored glass used for admitting light into the so-called dark-room, it is only necessary to place one of these instruments in the window to enable the operator to see whether the light entering his room is likely to injure the sensitive surfaces there exposed; thus, having ascertained by experience that his plates are fogged, or his paper injured, when the revolutions exceed, say, ten a minute, he will take care to draw down an extra blind when the revolutions approach that number. Still more useful will the radiometer be in the photographic gallery. Placing an instrument near the sitter at the commencement of the day's operations, it is found that, to obtain a good negative, the lens must be uncovered—not for a particular number of seconds—but during the time required for the radiometer to make, say, twenty revolutions. For the remainder of the day, therefore, assuming his chemicals not to vary, the operator need not trouble himself about the variation of light; all he has to do is to watch the radiometer and expose for twenty revolutions, and his negatives will be of the same quality,[6] although at one time it may have taken five minutes, and at another not ten seconds, to perform the allotted number.

I have long been experimenting in the endeavor to trace some connection between the movements of attraction and repulsion above alluded to and the action of gravitation in Cavendish's celebrated experiment. The investigation is not sufficiently advanced to justify further details, but I will give here an outline of one of the results.

I find that a heavy metallic mass, when brought near a delicately-suspended light ball, attracts or repels it under the following circumstances:

I. *When the ball is in air of ordinary density.*
☐ *a.* If the mass is *colder* than the ball, it *repels* the ball.
☐ *b.* If the mass is *hotter* than the ball, it *attracts* the ball.
II. *When the ball is in a vacuum.*
☐ *a.* If the mass is *colder* than the ball, it *attracts* the ball.
☐ *b.* If the mass is *hotter* than the ball, it *repels* the ball.

The density of the medium surrounding the ball, the material of which the ball is made, and a very slight difference between the temperatures of the mass and the ball, exert so strong an influence over the attractive and repulsive force, and it has been so difficult for me to eliminate all interfering actions of temperature, electricity, etc., that I have not yet been able to get distinct evidence of an independent force (not being of the nature of heat or light) urging the ball and the mass together.

Experiment has, however, shown me that, while the action is in one direction in dense air, and in the opposite direction in a vacuum, there is (as I have already pointed out in the experiments described in the commencement of this paper) an intermediate pressure at which differences of temperature appear to exert little or no interfering action. By experimenting at this critical pressure, and at the same time taking all the precautions which experience shows are necessary, it would seem that such an action as was obtained by Cavendish, Reich, and Baily, should be rendered evident.

It is not unlikely that in the experiments here recorded may be found the key of some as yet unsolved problems in celestial mechanics. In the sun's radiation passing through the *quasi* vacuum of space we have the radial repulsive force, possessing successive propagation, required to account for the changes of form in the lighter matter of comets and nebulae, and we may learn by that action, which is rapid and apparently fitful, to find the cause in those rapid bursts which take place in the central body of our system; but until we measure the force more exactly we shall be unable to say how much influence it may have in keeping the heavenly bodies at their respective distances.

So far as repulsion is concerned, we may argue from small things to great, from pieces of pith up to heavenly bodies; and we find that the repulsion shown between a cold and warm body will equally prevail, when for melting ice is substituted the cold surface of our atmospheric sea in space, for a lump of pith a celestial sphere, and for an artificial vacuum a stellar void.

Throughout the course of these investigations I have endeavored to remain unfettered by the hasty adoption of a theory, which, in the early stages of an inquiry, must almost of necessity be erroneous. Some minds are so constituted that they seem impelled to form a theory on the slightest experimental basis. There is then great danger of their becoming advocates, and unconsciously favoring facts which seem to prove their preconceived ideas, and neglecting others which might oppose their views. This is unfortunate, for the mind should always be free to exercise the judicial function, and give impartial weight to every phenomenon which is brought it. *Any* theory will account for *some* facts; but only the true explanation will satisfy *all* the conditions of the problem, and this cannot be said of any theory which has yet come to my mind.

My object at present is to ascertain facts, varying the conditions of each experiment so as to find out what are the necessary and what the accidental accompaniments of the phenomena. By working steadily in this manner, letting each group of experiments point out the direction for the next group, and following up as closely as possible, not only the main line of research, but also the little by-lanes which often lead to the most valuable results, after a time the facts will group themselves together and tell their own tale; the conditions under which the phenomena invariably occur will give the laws; and the theory will follow without much difficulty. The eloquent language of Sir Humphry Davy contains valuable advice, although in terms somewhat exaggerated. He says: "When I consider the variety of theories which may be formed on the slender foundation of one or two facts, I am convinced that it is the business of the true philosopher to avoid them altogether. It is more laborious to accumulate facts than to reason concerning them; but one good experiment is of more value than the ingenuity of a brain like Newton's."—*Quarterly Journal of Science.*

1.

- "On the Atomic Weight of Thallium," "Philosophical Transactions," 1873, vol. clxiii., p. 287.
- For a full description of this pump, with diagrams, see "Philosophical Transactions," 1873, vol. clxiii., p. 295.
- "Philosophical Transactions," 1874, vol clxiv., p. 501.
- This can be effected without interfering with the exhaustion.
- Some of the glass fibres used in these torsion-balances are so fine that when one end is held between the fingers the other portion floats about like a spider's thread, and frequently rises until it takes a vertical position.
- In this brief sketch I omit reference to the occasions in which the ultra-violet rays diminish in a greater proportion than the other rays.